# MeatEater's American History

The Mountain Men (1806-1840)

ISBN 978-1-300-52412-0
Steven Rinella
Copyright@2025

TABLE OF CONTENT

# CHAPTER 1

# INTRODUCTION

An Examination of the Mountain Men Era: Following the Louisiana Purchase

The era of the mountain men, which unfolded approximately between 1806 and 1840, arose in the aftermath of the Louisiana Purchase, a significant milestone in American history that effectively more than doubled the nation's territory. This significant acquisition from France in 1803 unveiled extensive, uncharted lands in the western region of the continent, fostering a spirit of boundless opportunity and exploration. The Rocky Mountains stood as a magnificent and formidable boundary, presenting an unexplored wilderness brimming with both opportunity and danger. Amidst this setting, a distinctive cohort of individuals—referred to as the mountain men—embarked on journeys into these territories, motivated by the promise of economic gain, the pursuit of personal liberty, and the captivating charm of the wild West.

The mountain men distinguished themselves as extraordinary figures among the ranks of pioneers. These individuals were engaged in the pursuits of trapping and hunting, embodying the spirit of adventure and exploration. They were driven not merely by the necessity of survival but also by a profound desire for autonomy and self-sufficiency. In contrast to settlers or farmers, these individuals flourished at the fringes of civilization, forging a life in some of the most harsh and inhospitable terrains on the continent. In this endeavor, they emerged as both representatives and emblems of westward expansion, capturing the essence of exploration that defined early 19th century America.

Exploring Prominent Personalities: Jim Bridger, Jedidiah Smith, Hugh Glass

Within the brotherhood of mountain men, specific individuals ascended to iconic prominence owing to their remarkable adventures and significant contributions to that period. Jim Bridger, for instance, gained fame for his profound understanding of the Rockies, his skill in traversing the wilderness, and his contributions as a guide for settlers and military expeditions. Bridger's existence exemplified the ingenuity and resilience necessary to thrive in the rugged terrain of the mountains.

Jedidiah Smith, a significant figure, was renowned for his relentless exploration of the American West. Fueled by an unquenchable thirst for knowledge and a resolute spirit, Smith emerged as one of the pioneering white men to traverse into California through the Sierra Nevada, embarking on an exploration of the Oregon Country. His travels not only broadened the understanding of geography but also showcased an extraordinary fortitude in confronting peril, including enduring a vicious bear attack that left him with deep scars yet unwavering in spirit.

Hugh Glass, a figure whose narrative has been enshrined in the annals of popular culture, is most notably recalled for his extraordinary endurance following a brutal encounter with a grizzly bear. Abandoned by those he once called allies, Glass persevered through immense hardship, crawling and stumbling across hundreds of miles in a relentless pursuit of safety, fueled by an indomitable spirit and a thirst for retribution. His narrative, both chilling and uplifting, embodies the fundamental battle for existence that characterized the experiences of numerous mountain men.

The contributions of these individuals, alongside countless unnamed and unsung mountain men, weave a rich tapestry of narratives that persist in captivating audiences and influencing perceptions of the American West.

Context: The Practice of Beaver Trapping and the Enchantment of the Rocky Mountains

The primary catalyst for the era of the mountain men was the beaver fur trade, a thriving enterprise propelled by the insatiable demand for felt hats in both Europe and America. The fur of the North American beaver, esteemed for its exceptional quality and adaptability, emerged as a highly coveted commodity. In the rugged expanse of the Rocky Mountains, intrepid trappers embarked on perilous journeys to harvest the plentiful beaver populations, engaging in the arduous tasks of capturing and skinning these creatures amidst the challenges of isolation and danger.

The appeal of the Rockies extended beyond mere economic prospects. For numerous individuals of the mountainous frontier, the area symbolized an opportunity to break free from societal limitations and adhere to their own principles. The expanse of the wilderness presented an unmatched sense of liberty and a chance to challenge oneself in the face of nature's formidable forces. The majestic mountains, characterized by their soaring peaks, profound valleys, and crystal-clear rivers, exuded an allure that attracted individuals in search of adventure and purpose within the wild expanse of the frontier.

The mountain men were not simply opportunists; they were pioneers who forged new paths, charted uncharted lands, and engaged with Native American tribes. In examining these interactions, it becomes evident that while certain exchanges were characterized by cooperation and mutual benefit, others were overshadowed by conflict and violence. The mountain men drew upon their deep understanding of the terrain, their connections with Indigenous communities, and their ingenuity to maneuver through the intricate challenges of wilderness existence.

Thesis: Mountain Men as Lasting Emblems of Autonomy, Tenacity, and Wild Essence in the Narrative of America

The mountain men transcended their roles as mere figures of the past; they emerged as cultural symbols, leaving a lasting imprint on the collective American psyche. Their existence, marked by autonomy, strength, and a profound bond with the environment, endures in the shared consciousness of the country. The individuals who journeyed into the Rocky Mountains in pursuit of their fortunes exemplified the adventurous spirit and self-sufficiency that have come to characterize the American identity.

Despite the brevity of their time, the impact of the mountain men remains significant. The narratives, rich with both valor and struggle, embody the intricacies of human experience and the profound influence of the untamed landscape. In forging their existence in the Rockies, they played a pivotal role in the territorial growth of the United States while simultaneously creating an enduring symbol of self-reliance and the wild essence of the frontier. In contemporary discourse, the mountain men are revered as emblematic figures of a time when the West represented both promise and peril, a realm where aspirations and the instinct for survival were inextricably linked.

# CHAPTER 2

## Historical Background

The Louisiana Purchase of 1803: A Catalyst for Westward Expansion

The Louisiana Purchase, completed in 1803 during the presidency of Thomas Jefferson, marked a pivotal moment in the narrative of American history. The acquisition of roughly 828,000 square miles from France for the sum of $15 million resulted in a significant expansion of the United States, effectively doubling its size and pushing its western boundary from the Mississippi River to the Rocky Mountains. This significant land agreement not only settled territorial conflicts with European nations but also ignited the aspirations of a youthful and swiftly expanding country. The recently obtained territories embodied vast possibilities—rich earth for farming, pristine natural assets, and an apparently limitless expanse ready for discovery.

Central to the Louisiana Purchase was the inquiry into the nature of the uncharted territories that lay before them. In response to this need, Jefferson initiated expeditions such as the renowned Lewis and Clark journey (1804–1806), which meticulously charted pathways across the West while recording its geography, flora, fauna, and the cultures of its indigenous peoples. The initial explorations unveiled the grandeur of the Rocky Mountains, which stood as a formidable natural obstacle and a wellspring of intrigue and difficulty for those who would come to settle in the region.

The acquisition carried additional noteworthy consequences as well. This phenomenon ignited a fervor for expansionism—though the phrase itself would not emerge until the mid-19th century—that led many Americans to perceive it as their ordained responsibility to move westward. The majestic Rockies, characterized by their lofty summits and untouched landscapes, emerged as an emblem of possibility and exploration. For the Mountain Men, these formidable terrains offered not merely financial opportunity but also the prospect of forging a life marked by exceptional autonomy and strength of spirit.

The Beaver Fur Trade: Economic Demand and Its Influence on Exploration Initiatives

The Mountain Men were predominantly motivated by a singular economic impetus: the beaver fur trade. During the early 19th century, there was a remarkable surge in the global demand for beaver pelts, driven by a burgeoning fascination with felt hats in both Europe and the United States. The esteemed qualities of beaver fur—its remarkable durability, exceptional warmth, and versatile nature—established it as a fundamental element within the realm of fashion. The pelts were transformed into premium felt, a material that came to represent both status and sophistication in the realm of fashionable headwear for the upper class.

The surge in demand fostered a thriving market for fur traders, transforming the Rocky Mountains into a focal point for beaver populations. The myriad streams and rivers of the region created a perfect environment for these hardworking aquatic mammals, transforming the Rockies into a true treasure trove for those daring enough to venture into the wild to capture them. Resourceful pioneers—typically youthful, intrepid, and driven by the allure of prosperity—migrated to the area, aspiring to achieve success in the fur commerce.

The economic framework of the beaver fur trade was distinctive and significantly influenced the way of life of the Mountain Men. Prominent fur trading enterprises, including the American Fur Company, initiated by John Jacob Astor, alongside the Rocky Mountain Fur Company, developed frameworks to facilitate trapping endeavors. The companies engaged the services of Mountain Men as trappers, frequently under contractual agreements, or functioned through the rendezvous system. The yearly gathering served as a pivotal occasion: trappers, traders, and Indigenous peoples convened at designated sites to barter furs for provisions, armaments, and various other items. This occasion transcended mere commerce, evolving into a gathering rich with fellowship, narrative exchange, and joyous celebration.

The beaver fur trade facilitated direct interactions between the Mountain Men and various Native American tribes, who had established traditions of fur trapping over many generations. Some relationships exhibited cooperation, as Native Americans frequently took on roles as guides, trading partners, or even family members through intermarriage, while others were marked by significant tension and conflict. The struggle for resources and land unavoidably resulted in conflicts, introducing yet another dimension of peril to the existence of the Mountain Men.

The beaver fur trade, though it began with great promise, ultimately faced certain constraints. As the years progressed, the rigorous trapping practices resulted in a notable decrease in beaver populations, a situation further intensified by evolving fashion trends that diminished the demand for beaver felt. By the mid-19th century, the commerce that had characterized the era of the Mountain Men was swiftly declining, compelling many of them to either adjust to new realities or fade into obscurity.
Geographic Focus: The Rockies as a Wild and Uncharted Territory

The Rocky Mountains, extending more than 3,000 miles from Canada to New Mexico, served as the essential spine of the American frontier throughout the era of the Mountain Men. This expansive and formidable range is marked by its lofty summits, thick woodlands, and a complex system of rivers and valleys. For the Mountain Men, the Rockies represented not only a formidable challenge to overcome but also a wellspring of boundless opportunity.

The terrain of the Rockies exhibited a remarkable diversity, coupled with an imposing nature. In the northern regions, the landscape was characterized by valleys shaped by glacial activity and vast plateaus, whereas the central and southern sections of the Rockies displayed sharp peaks and slender canyons. The natural beauty of the region concealed its perils: severe winters, capricious weather, and the looming threat of avalanches rendered survival an ongoing challenge. For those lacking preparation, the Rockies could prove to be as perilous as they were magnificent.

It was this wild and unrestrained landscape that captivated the Mountain Men. The Rockies provided an unparalleled sense of freedom and self-sufficiency, distinct from the constraints found in more settled regions. In this setting, they sustained themselves by utilizing the resources around them, engaging in hunting and trapping, while demonstrating a profound connection with their surroundings. The intricate network of rivers and streams that traversed the region served as vital conduits for exploration and trapping, granting entry to otherwise unreachable territories.

The Rocky Mountains served as a habitat for a variety of ecosystems and a rich array of wildlife. In addition to beavers, the area was abundant with elk, deer, bighorn sheep, and grizzly bears, among various other species. In the intricate tapestry of survival among the Mountain Men, each animal played a pivotal role: beavers were sought after for their prized pelts, while the larger game offered sustenance, apparel, and essential tools. Nevertheless, the existence of formidable predators such as grizzlies and mountain lions introduced an additional dimension of peril to their everyday existence.

The geographical isolation of the Rockies significantly influenced the cultural development of the Mountain Men. Isolated from the familiar comforts and limitations of society, they cultivated a distinctive bond of fellowship and a strong sense of self-reliance. Their deep understanding of the terrain, cultivated over many years, garnered them admiration and recognition as some of the most proficient outdoorsmen of their era.

Alongside its inherent natural challenges, the Rockies presented a rich tapestry of cultural exchanges. This area served as a dwelling for a variety of Indigenous tribes, such as the Crow, Blackfeet, and Shoshone. For the Mountain Men, traversing these cultural landscapes held equal significance to maneuvering through the physical terrain. Partnerships and commercial ties with Indigenous tribes were essential for endurance, whereas confrontations frequently escalated into violence.

By the conclusion of the Mountain Men's era, the Rockies had experienced a significant transformation. What was once a vast and untamed expanse has gradually transformed into a well-documented and inhabited landscape. The paths forged by these initial adventurers and fur traders transformed into the corridors for pioneers and settlers journeying westward, indelibly reshaping the geography and heritage of the area.

# CHAPTER 3

## Life in the Wilderness

A. Sustenance from the Earth

Enduring the challenges of the isolated wilderness of the Rocky Mountains was a remarkable achievement. The mountain men inhabited some of the most formidable landscapes on the continent, characterized by rugged terrain, severe weather conditions, and limited resources. To ensure their survival, they drew upon their profound understanding of the terrain and refined abilities in hunting, fishing, and foraging. Their sustenance was largely derived from the offerings of the wilderness, including elk, bison, and deer, with the occasional inclusion of smaller game such as rabbits and squirrels. In periods of dire need, they turned to consuming roots, berries, and, at times, their own horses or mules.

Their provisions were minimal, consisting solely of firearms, traps, knives, and a handful of tools necessary for cooking and the upkeep of their gear. Their garments were frequently crafted from animal hides, offering both resilience and warmth in response to the severe conditions of the mountainous environment. The experience of residing in such seclusion necessitated a significant level of self-sufficiency, given the absence of stores, medical assistance, or any other external resources.

The interactions between the mountain men and Native peoples represented a vital component of their endurance and existence. The Crow, Blackfoot, Shoshone, and Flathead tribes engaged in a complex interplay of competition for resources, while simultaneously exploring opportunities for alliances and trade partnerships. Certain individuals who ventured into the mountains formed advantageous partnerships with Indigenous groups, trading items such as firearms, metal implements, and beads in return for furs, sustenance, and navigation assistance through the landscape. Some individuals entered into marriages with Indigenous women, thereby reinforcing their connections to local tribes and fostering collaboration.

Nevertheless, these associations were fraught with discord. Conflicts frequently arose from cultural misunderstandings, competition for resources, and the intrusion of mountain men into revered territories, resulting in violent confrontations. The capacity to maneuver through these exchanges—whether collaborative or confrontational—was an essential skill for enduring in the wild.

The capacity for endurance stood as a hallmark of the lives led by the mountain men. Their existence was marked by unyielding environmental adversities, encompassing frigid winters, scorching summers, and the ever-present peril of natural calamities such as avalanches and flash floods. Interactions with wildlife introduced an additional dimension of peril, with grizzly bear attacks standing out as a particularly infamous threat. Enduring such an encounter demanded not merely physical prowess but also swift intellect and an extraordinary resolve to survive. Accounts of individuals such as Hugh Glass, who endured a bear attack and was abandoned, highlight the remarkable tenacity of these figures.

B. The Economy of Beaver Trapping

The fur trade stood as the central pillar of the mountain men era, with beaver pelts being the fundamental element of this economic endeavor. The allure of beaver fur was largely propelled by the fashion trends of Europe and America, particularly the esteemed felt hats crafted from beaver pelts that captured the admiration of society. The burgeoning demand fostered a profitable arena for fur traders, drawing enterprising souls to the Rocky Mountains, a region where beaver populations flourished.

Individuals who ventured into the mountains employed distinct tools and methods for capturing beavers, frequently situating their traps along waterways where these creatures were recognized for constructing their homes and barriers. The steel traps were expertly baited with castoreum, a substance obtained from the scent glands of beavers, effectively attracting the animals. Upon capture, the beavers met their demise, their skins meticulously removed and prepared for the purposes of trade. The value of a single prime-quality beaver pelt could command a significant price, rendering the challenges and dangers of trapping a worthwhile endeavor for numerous individuals.

The fur trade revolved around yearly gatherings, where trappers, traders, and Indigenous communities convened to barter goods, share tales, and disseminate knowledge. These assemblies, frequently convened in secluded areas, served as both social and economic occasions. In the rugged expanse of the wilderness, mountain men engaged in the exchange of their furs for essential supplies such as gunpowder, ammunition, tobacco, and alcohol, alongside the indulgences that eluded them in their remote surroundings. The gathering also provided an opportunity to pause, rekindle relationships with former companions, and ready oneself for the upcoming trapping season.

Although it yielded significant profits, the beaver trapping economy was fundamentally unsustainable. The excessive trapping of beavers resulted in a significant decrease in their populations, while shifting fashion trends in Europe ultimately diminished the demand for fur. Consequently, the economic basis of the mountain men's lifestyle started to decline, compelling numerous individuals to adjust by assuming positions as guides, scouts, and explorers during the period of westward expansion.

C. Profound Understanding of the Rockies

The survival and success of the mountain men hinged on their exceptional understanding of the Rocky Mountains, a terrain that was both stunning and perilous. They were pioneers in the exploration and mapping of this expansive area, playing a crucial role in enhancing the comprehension of its geographical features.

Traversing the Rockies demanded a profound comprehension of the landscape, climatic conditions, and available natural resources. The mountain men developed a keen ability to discern landmarks, pinpoint secure pathways, and find water sources in a landscape where even the slightest error could prove deadly. This understanding was frequently gained over many years of experience and, in numerous instances, through engagements with Indigenous communities, who had coexisted with the land for countless generations.

The proficiency of the mountain men in the wild encompassed their remarkable capacity to adjust and create new solutions. They fashioned implements and dwellings from the resources at hand, devised creative responses to unforeseen obstacles, and formulated approaches to address the physical and mental challenges posed by their surroundings. Their ingenuity was clearly demonstrated through their capacity to navigate perilous mountain routes, withstand extended stretches of solitude, and confront dangers posed by both humanity and the natural world.

The profound understanding of the Rockies rendered the mountain men essential to the westward expansion of the United States. As pioneers ventured into the western territories, these trappers and explorers took on the role of guides, directing wagon trains, surveying the land, and forging new pathways. Their impact on the exploration and settlement of the American West is of immense significance.

# CHAPTER 4

## The Brotherhood of Mountain Men

A. A Limited Brotherhood: Comprising Merely Several Hundred at Any Given Moment

The mountain men established a distinctive and secluded community, commonly described as a "brotherhood" because of their collective lifestyle and interdependence in confronting significant challenges. Although their influence on the exploration and development of the American West was profound, the actual figures were quite modest. Throughout history, the brotherhood of mountain men seldom numbered more than a few hundred individuals at any one moment. Throughout the period from 1806 to 1840, it is believed that a maximum of 3,000 individuals engaged in the pursuits of trapping and exploring the untamed landscapes of the Rocky Mountains.

This exclusivity arose from the arduous conditions of the mountain man's life. The vocation demanded a unique blend of physical prowess, mental fortitude, and an unwavering quest for liberty and exploration. The rigors of trapping and enduring the harsh conditions of the Rockies were not a fit for all individuals. Individuals who chose to enlist were frequently motivated by a desire for autonomy, a sense of exploration, or the enticing prospects of the profitable fur trade. Numerous individuals were youthful males, keen to challenge their capabilities and demonstrate their value in an environment that demanded unwavering resolve.

The limited population of mountain men cultivated a profound sense of fellowship and collective identity within their ranks. While frequently working independently or in small groups for trapping, they united during significant occasions like the annual rendezvous, where they engaged in trade, shared tales, and strengthened the ties of their distinctive camaraderie. The social dimension of their existence, albeit sporadic, played a crucial role in their survival and psychological health, offering a unique sense of belonging in a predominantly isolated life.

B. Characteristics and Values: Autonomy, Resilience, and Self-Sufficiency

The mountain men embodied a distinct set of traits and principles that distinguished them from their fellow pioneers and settlers, elevating them to iconic status in the narrative of American history. At the forefront of these characteristics were self-sufficiency, determination, and an unwavering spirit of autonomy.

The quality of self-reliance stood as a paramount trait for those who ventured into the mountains. In the untamed expanse, they relied solely on their own resources and resilience. They were required to master a diverse array of skills, encompassing hunting and trapping, as well as navigation and the construction of shelters. Their continued existence frequently depended on their capacity to adjust to unforeseen circumstances, including abrupt shifts in climate, interactions with untamed creatures, or confrontations with fellow humans. For the mountain men, self-reliance transcended mere virtue; it was an essential requirement, woven into the very fabric of their existence.

The mountain men were characterized by a remarkable blend of courage, resolve, and unwavering determination, which can be described as grit. The obstacles encountered on a daily basis would have overwhelmed those with less resolve. Through the harsh realities of prolonged winters in rudimentary shelters and the struggle to overcome injuries and ailments without the benefit of medical assistance, their existence was a relentless examination of both physical strength and mental resilience. The accounts of individuals such as Jedidiah Smith, who endured a bear attack resulting in a torn scalp and exposed ribs, yet persisted in his explorations, underscore the remarkable resilience exhibited by these men.

Ultimately, the essence of the mountain men's identity was rooted in their independence. In contrast to those who endeavored to establish communities or those who aimed to till the soil, the mountain men flourished in solitude. They cherished the liberty to navigate their own journeys, liberated from the confines of societal norms or anticipations. This strong sense of autonomy led to both admiration and misunderstanding among their peers, who frequently perceived them as rugged individualists or, at times, as outlaws. In contemporary discourse, this characteristic is revered as an integral aspect of the lasting narrative surrounding the American West.

C. Peril and Demise: A Tenth of Fatalities Resulting from Violence in the Rocky Mountains

The existence of a mountain man was laden with peril, and the rate of mortality within their ranks was alarmingly elevated. It is believed that one in ten mountain men encountered a violent demise in the wilderness—a figure that highlights the treacherous reality of their lives. The reasons behind these fatalities were as diverse as they were horrific, encompassing everything from assaults by animals and exposure to brutal confrontations with fellow humans.

Interactions with wildlife represented one of the most pressing dangers faced by the mountain men. Grizzly bears, notably, posed a persistent threat. These formidable creatures, capable of exceeding 500 kilograms in weight and reaching heights of nearly 3 meters when on their hind legs, exhibited both territorial behavior and aggression. A considerable number of mountain men succumbed to bear attacks, whereas others, such as Hugh Glass, managed to endure the harrowing experience with great difficulty. Among the various perils faced were wolves, mountain lions, and venomous snakes, alongside the risks associated with horse-related accidents or the treacherous possibility of falling from cliffs.

The unforgiving landscape of the Rockies has, throughout history, taken a significant toll on human life. Frigid temperatures, unexpected snowstorms, and the threat of avalanches were prevalent dangers, particularly for those who dared to explore elevated terrains. The perils of frostbite, hypothermia, and dehydration loomed large, alongside the ever-present threat of injuries stemming from falls or mishaps during the pursuits of trapping and hunting.

The presence of human conflict introduced an additional dimension of peril. The mountain men frequently encountered conflicts regarding trapping territories, engaging in disputes not only with fellow trappers but also with Native American tribes. In the complex tapestry of history, certain mountain men forged amicable connections with Indigenous communities, while others found themselves embroiled in violent clashes. Conflicts often arose from cultural misunderstandings, competition for resources, and the intrusion of trappers into revered territories, resulting in tragic confrontations.

In the face of these dangers, the mountain men embraced the prospect of death as an intrinsic aspect of their selected existence. Their readiness to confront such perils reflects an exceptional level of bravery and resolve. For these individuals, the allure of the untamed wilderness and the spirit of adventure outweighed any cost, even if it led to a premature and brutal conclusion.

# CHAPTER 5

## Heroic and Horrifying Feats

A. Notable Achievements and Interactions

The mountain men are commemorated not solely for their adeptness and resilience in the wilderness but also for their remarkable narratives, numerous of which intertwine the boundaries of fact and folklore. The narratives of bravery, resilience, and sheer human determination have enthralled countless generations, depicting the mountain men as extraordinary individuals beyond the ordinary realm. Among these figures, Jim Bridger, Jedidiah Smith, and Hugh Glass are distinguished by their extraordinary adventures and unwavering determination.

1. The Enduring Tales of Jim Bridger

Jim Bridger stands as a prominent figure among the mountain men, renowned for his remarkable skills in navigating and enduring the challenges of the wilderness. Born in 1804, Bridger embarked on his journey as a trapper early in life, dedicating many years to the exploration of the Rocky Mountains. His remarkable tales of endurance established him as one of the most ingenious and informed pioneers of his era.

Among Bridger's notable exploits is the harrowing tale of his narrow escape from a perilous encounter with the Blackfoot tribe. During a trapping expedition, Bridger and his companions encountered an ambush, compelling them to navigate back across perilous terrain. Bridger, abandoned to his own devices, journeyed across vast distances in solitude, drawing upon his understanding of the terrain to evade capture and sustain himself with scant provisions. This remarkable achievement showcased his extraordinary capacity to adjust and endure in perilous circumstances.

Bridger was renowned for his sharp wit and captivating storytelling, frequently weaving tales of his adventures that artfully combined elements of reality and imagination. The narratives surrounding his encounters with geysers and hot springs—subsequently recognized as elements of modern Yellowstone—enhanced his enigmatic persona and solidified his status as one of the most vibrant characters of the mountain men period.

2. The Exploration and Resilience of Jedidiah Smith

Jedidiah Smith stands out as a figure of remarkable courage and resolve in the exploration of the American West. In contrast to numerous individuals of his time, Smith held strong religious beliefs and refrained from consuming alcohol, which led to him being affectionately referred to as The Bible Toter. His unwavering determination and dedication to discovery enabled him to achieve remarkable accomplishments that appeared nearly beyond human capability.

Among the notable events in Smith's life is the harrowing incident in which he was viciously attacked by a grizzly bear while on an expedition in the year 1823. The bear launched its assault abruptly, inflicting severe wounds that stripped away Smith's scalp and laid bare his ribs. In a remarkable turn of events, Smith managed to endure the assault and, with the assistance of his comrades, meticulously reattached his scalp utilizing a needle and thread. In the face of severe injuries, Smith pressed on with his journey merely days later, showcasing remarkable resilience and determination.

Smith's explorations were remarkably significant. He was one of the earliest individuals from outside the Indigenous communities to traverse the Sierra Nevada into California and to chart the previously uncharted areas of the region. His detailed journals and maps offered crucial insights into the American West, playing a vital role in shaping the nation's comprehension of its expansive frontier. Smith's unwavering quest for knowledge and his capacity to withstand extraordinary challenges established him as a remarkable figure of the mountain men era.

3. The Grizzly Bear Encounter of Hugh Glass and His Quest for Retribution

The tale of Hugh Glass stands out as one of the most renowned among the legends of the mountain men, celebrated in literature, oral traditions, and the cinematic portrayal in The Revenant. Born circa 1783, Glass later took on the role of a fur trapper, swiftly establishing a reputation for his remarkable tenacity and exceptional survival skills. Nevertheless, it was his harrowing encounter with a grizzly bear that solidified his significance in the annals of time.

In the year 1823, during an expedition aimed at scouting for fur trapping opportunities, Glass encountered a ferocious grizzly bear, resulting in severe and life-threatening injuries. The bear inflicted severe injuries upon him, resulting in a broken leg, deep lacerations, and exposed ribs. Convinced of his impending demise, his companions—John Fitzgerald and a youthful Jim Bridger—were assigned the solemn duty of remaining behind to tend to his needs. Rather, they forsook him, seizing his rifle and provisions, and left him to perish.

Defying the most formidable challenges, Glass emerged unscathed. Motivated by an indomitable spirit and a quest for retribution, he traversed over 200 miles, crawling and stumbling, in his pursuit of safety. Throughout his journey, he faced intense suffering, relied on foraged berries and uncooked flesh for sustenance, and bravely defended himself against wolves. His resolve to face those who had wronged him propelled his remarkable odyssey.

Upon reaching civilization, Glass confronted Fitzgerald and Bridger, yet he ultimately decided to spare their lives. The narrative serves as a profound reflection of human resilience and the determination to persevere, emerging as one of the most emblematic accounts from the mountain men era, capturing the harsh truths and the valorous essence of existence in the untamed wilderness.

### B. The Idealized Versus the Harsh Truths of Their Existence

The narratives surrounding Jim Bridger, Jedidiah Smith, Hugh Glass, and their contemporaries have been embellished throughout history, influencing the common view of these mountain men as intrepid explorers and resilient lone figures. Though these accounts convey the thrill and valor of their endeavors, they frequently mask the stark and relentless truths of their existence.

The idealized portrayal of the mountain men presents them as extraordinary individuals who flourished in the untamed wilderness, driven by unwavering resolve and tenacity. Their narratives are rich with adventure, peril, and victory, portraying them as embodiments of the wild essence of the American West. This depiction has been sustained through various forms of literature, art, and film, enriching the narrative of the frontier and the concept of rugged individualism.

Nevertheless, the intricacies of their existence were considerably more nuanced and frequently harsh. The wilderness presented itself not as a picturesque setting, but rather as a relentless and unyielding landscape that required unwavering attention and toil. The mountain men encountered a series of formidable obstacles, such as harsh climatic conditions, limited supplies, and the constant peril of injury or demise. A significant number endured the ravages of malnutrition, disease, and debilitating injuries, while others fell victim to violent encounters with both wildlife and fellow humans.

The complexities of their interactions with Indigenous communities introduce significant moral and ethical considerations that challenge the idealized accounts often presented. In examining the interactions between mountain men and Indigenous communities, it is evident that while some individuals fostered respectful relationships, others engaged in violent confrontations and sought to exploit these groups for their own benefit. The fur trade, a significant force behind their endeavors, played a crucial role in the depletion of natural resources and the disruption of Native American livelihoods.

Indeed, the existence of the mountain men was characterized equally by trials and tribulations as it was by valor and exploration. The narratives, though uplifting, uncover the intricate layers and paradoxes of human nature when confronted with severe challenges. Through a careful exploration of the idealized and harsh realities of their existence, we achieve a deeper comprehension of their impact and the time period they significantly shaped.

# CHAPTER 6

## The Rise of the Mountain Men: Peak Era

The demand for beaver fur catalyzed significant economic prosperity.

The Mountain Men arose in a distinctive economic period in both American and global history, characterized by the peak demand for beaver pelts. This demand was not a mere coincidence; it emerged from a confluence of economic, social, and industrial influences that elevated the beaver fur trade to one of the most profitable enterprises of the early 19th century. The resilient trappers flourished within this economic framework, establishing themselves as the cornerstone of the fur trade and experiencing a notable era of prosperity.

During the early 19th century, beaver pelts emerged as the sought-after raw material for the production of superior felt, especially in the crafting of hats. Felt hats, once the hallmark of Europe's upper echelons, served as emblems of prestige, sophistication, and affluence. The crafting of these hats was fundamentally dependent on the thick, water-resistant undercoat of beaver fur, rendering it a crucial resource for milliners and the fashion sector. This development paralleled the decline of European beaver populations, a consequence of excessive hunting, compelling traders to seek opportunities in North America, where abundant beaver-rich regions were still available.

For fur enterprises such as the American Fur Company, Hudson's Bay Company, and Rocky Mountain Fur Company, the Rocky Mountains symbolized the final significant frontier of the beaver trade. The formation of these companies led to the establishment of an intricate network comprising trappers, traders, and supply chains within the industry. The Mountain Men served an essential function as the "boots on the ground," journeying into the untamed wilderness to procure the highly sought-after pelts.

The yearly gathering system emerged as a defining feature of this economic framework. Arranged by fur companies, these assemblies functioned as trading fairs where trappers traded their pelts for vital supplies, tools, and luxuries provided by company representatives. Rendezvous served as vital centers of commerce and social interaction, providing Mountain Men with opportunities to exchange tales, commemorate their experiences, and strategize for future adventures. It was during these gatherings that the economic contributions of the Mountain Men to the fur trade became distinctly apparent, as they provided the companies with the essential raw materials that fueled the industry's progress.

At the zenith of the fur trade, the Mountain Men experienced a period of notable affluence. Although the untamed landscape presented numerous perils, the monetary gains for those who triumphed were substantial. For numerous individuals engaged in this endeavor, the beaver fur trade offered a profound sense of purpose and achievement, allowing them to attain economic independence in manners that were frequently inaccessible in the more established regions of the United States.

Nevertheless, the prosperity derived from the beaver trade was accompanied by considerable repercussions. As the years progressed, the relentless pursuit of beavers through trapping resulted in a significant decrease in their numbers, thereby disrupting ecological balance and undermining the viability of the trade. Furthermore, the evolving fashion trends in Europe and America started to move away from beaver felt, leading to a decline in demand. The interplay of these elements would ultimately lead to the waning of the Mountain Men's era; however, at its zenith, the fur trade served as a catalyst that propelled these intrepid explorers into the chronicles of American history.

Their Distinct Contribution as Trailblazers of the West

In addition to their function as economic agents, the Mountain Men established a legacy as pioneers and pathfinders, significantly influencing the narrative and lore of the American West. Their lives and exploits were characterized by a profound spirit of exploration and adventure, echoing the ideals of independence, self-reliance, and resilience. The Mountain Men, in numerous respects, epitomized the fierce independence that would ultimately shape the essence of the American spirit.

The Mountain Men embarked on journeys into vast, largely uncharted territories, positioning themselves among the earliest non-Indigenous explorers of the Rocky Mountains and its adjacent areas. Their pursuits for streams abundant with beavers led them into secluded valleys, thick woodlands, and perilous mountain routes. Through their efforts, they broadened the understanding of the geographical landscape of the United States, offering essential perspectives on the terrain, natural resources, and possible pathways for westward expansion.

Individuals such as Jedidiah Smith exemplified the significant contributions of the Mountain Men as pioneers in exploration. Smith's audacious journeys led him through the Mojave Desert, over the Sierra Nevada, and along the Pacific Coast, positioning him among the earliest Americans to navigate these formidable terrains. The meticulous records and cartographic representations he created became essential resources for subsequent settlers and pioneers, underscoring the significant role played by the Mountain Men in the country's expansion to the west.

The survival of the Mountain Men hinged on their exceptional understanding of the terrain. They skillfully navigated the complexities of survival in the wild, adeptly employing the bounty of nature for sustenance, habitation, and attire. This close connection with the environment distinguished them from the settlers who came after, many of whom depended on the Mountain Men's expertise to traverse the unknown landscape.

Beyond their understanding of the terrain, the Mountain Men served as cultural bridges, interacting with the Indigenous tribes that resided in the Western regions. The interactions between the Mountain Men and Indigenous peoples were diverse, encompassing alliances and trade as well as instances of conflict and violence. In many cases, the Mountain Men depended on Native guides, technologies, and survival strategies for their endeavors. Through these interactions, they played a significant role in a multifaceted cultural exchange that influenced the narrative of the frontier.

The contributions of the Mountain Men as trailblazers reached far beyond the confines of mere exploration. They played a crucial role in the establishment of trails and routes that would eventually serve as vital corridors for settlers, traders, and travelers journeying westward. The Oregon Trail, for instance, was significantly influenced by the routes established by legendary figures such as Jim Bridger and Kit Carson. The establishment of these routes enabled the movement of countless settlers, reshaping the West from a vast, unexplored wilderness into a thriving tapestry of interconnected communities and economic systems.

The adventures of the Mountain Men significantly contributed to the narrative surrounding the American West. Their remarkable lives and bold exploits captivated the minds of writers, artists, and storytellers, who enshrined them as iconic figures in the annals of history. The narratives surrounding Hugh Glass's remarkable endurance following a grizzly bear assault and Jim Bridger's confrontations with unfriendly tribes have transcended mere storytelling, solidifying the Mountain Men's significance in the collective memory of the nation.

Nevertheless, this idealized portrayal frequently masked the stark truths of their existence. The Mountain Men encountered relentless peril, persevering through harsh conditions, solitude, and the looming specter of violence. Their ability to withstand such hardships emerged as a hallmark of their existence, yet it also highlighted the persistent difficulties of navigating a reality in which survival was perpetually uncertain.

At the height of their influence, the Mountain Men stood at the forefront of

American expansion, extending the limits of the known world and establishing the foundations for the settlement and growth of the West. Their contributions to exploration, commerce, and cultural exchange have profoundly influenced the narrative of the United States, even as their traditional way of life waned with the diminishing fur trade and the advance of civilization.

# CHAPTER 7

## Decline of the Mountain Men Era

A. Economic and Ecological Considerations

The waning of the mountain men era can be attributed primarily to a confluence of economic and ecological influences that eroded the essential basis of their existence: the beaver fur trade.

1. Decrease in the Beaver Population

By the mid-1830s, the once-thriving beaver populations in the Rocky Mountains and adjacent areas faced a significant downturn. The relentless trapping efforts by mountain men over the decades inflicted severe damage on the species, resulting in significant overexploitation. Trappers, driven by the allure of quick financial gain, frequently overlooked the implications for the enduring viability of the beaver population. Waterways that were once abundant with beavers gradually became desolate, compelling trappers to explore further into uncharted and isolated regions in pursuit of fresh resources.

The ecological ramifications of this excessive trapping reached far beyond the mere vanishing of beavers. Beavers serve an essential function in the preservation of vibrant ecosystems, as their dams foster the formation of wetlands that nurture a wide array of plant and animal species. The decline in beaver populations has significantly altered the ecological equilibrium in numerous areas, resulting in modifications to water flow, the disappearance of habitats, and the gradual deterioration of the landscape. The circumstances faced by the mountain men grew increasingly arduous, as the very environment upon which they depended gradually transformed into a less welcoming landscape.

## 2. Transformations in Worldwide Fashion Trends

As the decline of beaver populations posed an ecological obstacle to the fur trade, a parallel and equally pressing challenge emerged in the economic realm: the evolving preferences of consumers in Europe and America. During the early 19th century, beaver fur emerged as a coveted commodity, especially for the production of felt hats, which held a prominent place in European fashion. Nevertheless, as the

1830s drew to a close, a transformation in these trends became evident.

The emergence of silk, characterized by its lighter weight, affordability, and growing accessibility through advancements in manufacturing and trade, marked a significant shift as it began to supplant beaver fur as the favored material for hats. With the decline in demand for beaver pelts, the profits that had previously supported the fur trade also began to diminish. The European markets, once the foremost proponents of the industry, have largely turned away from beaver fur, resulting in American trappers facing diminished buyers and a decline in their earnings.

The interplay of ecological scarcity and diminishing economic demand ultimately marked the decline of the beaver fur trade, leaving the mountain men bereft of their main source of income. In the absence of a strong demand for pelts, numerous trappers found themselves compelled to reevaluate their positions within the wilderness economy, marking the onset of a significant era of change and unpredictability.

B. The Dismantling of Their Cultural Existence

With the decline of the beaver fur trade, the customary existence that characterized the mountain men also began to fade away. As certain individuals navigated the shifting landscape, others faced growing marginalization amid the unyielding advance of westward expansion and the evolution of the American frontier.

1. Shift from Trapping to Guiding and Trading Functions

In response to the diminishing prospects within the fur trade, a significant number of mountain men adapted by assuming new positions as guides, scouts, and traders. Their profound understanding of the Rocky Mountains and the adjacent wilderness rendered them essential allies to settlers, military expeditions, and explorers embarking on westward journeys. Figures such as Jim Bridger and Kit Carson emerged as iconic personalities in this role, guiding wagon trains along the Oregon Trail, contributing to the cartography of unexplored lands, and enabling the transportation of goods and individuals.

In their capacities as navigators and pathfinders, the former mountain men significantly contributed to the larger narrative of westward expansion, assisting settlers in traversing the perilous landscapes and establishing new pathways to California, Oregon, and various other frontier locations. Nevertheless, this shift encountered numerous obstacles. Guiding represented a significant departure from the autonomy and self-sufficiency associated with trapping, as it frequently necessitated a collaborative relationship with settlers and government officials. For certain individuals of the mountainous frontier, the relinquishment of independence associated with these emerging positions proved to be a challenging reality to embrace.

Some individuals engaged in commerce, setting up trading posts to provide settlers and Indigenous communities with goods in return for food, furs, and various resources. These trading posts, frequently positioned along significant trails and waterways, emerged as centers of commerce and culture within the developing frontier landscape. Although trading enabled certain former trappers to preserve a link to the wilderness, it simultaneously signified a shift away from the solitary and adventurous existence they had previously experienced.

## 2. The Incorporation or Displacement Resulting from the Growth of American Settlements

With the westward expansion of American settlements, the once untamed wilderness that had belonged to the mountain men began its gradual metamorphosis into bustling towns, fertile farms, and essential infrastructure. The vast and wild expanse of land gradually succumbed to regulation and division, diminishing the space available for the adventurous ways of the mountain men. A multitude became captivated by the expanding tapestry of American society, assimilating into communities as laborers, ranchers, or merchants.

For certain individuals, this integration heralded fresh opportunities, yet for others, it signified the demise of a cherished way of life. The essence of freedom and exploration that characterized the mountain men appeared at odds with the organized and established realm of urban life and farming. Consequently, numerous mountain men found it challenging to adjust, with some opting to withdraw deeper into the wilderness in an effort to maintain a semblance of the liberty they had previously experienced.

In the course of American expansion, the encroachment of settlements frequently resulted in the profound disruption or annihilation of Native peoples' lands and cultures, a consequence of both the surge of settlers and the prevailing policies enacted by the U.S. government. The mountain men, who previously navigated the fringes of both Indigenous and settler communities, became ensnared in the midst of this significant change. Some aligned themselves with the settlers, facilitating the displacement of Indigenous tribes, whereas others lamented the loss of their Native allies and the shared territories that had once united them.

The Conclusion of a Significant Period

By the 1840s, the era of the mountain men had predominantly reached its conclusion. The elements that had upheld their existence—plentiful wildlife, a thriving market for furs, and the seclusion of the frontier—had ceased to exist. Even as their roles transformed or diminished, the mountain men imparted a lasting influence on the narrative and heritage of the American West.

Their efforts in exploration, commerce, and the movement westward established the foundation for the colonization of the frontier. The narratives, encompassing both valor and warning, wove themselves into the fabric of the nation's folklore, igniting inspiration across generations through their tenacity, autonomy, and indomitable essence. Although the mountain men have disappeared from the wilderness, their legacy endures, serving as a reminder of a time when the West embodied both opportunity and challenge, freedom and peril.

# CHAPTER 8

## Legacy of the Mountain Men

A. Influence on the American West

The mountain men were instrumental in influencing the formative years of the American West's narrative. Despite their limited numbers, the impact they had on the exploration, settlement, and economic advancement of the region was nothing short of monumental. Their efforts have left an indelible mark on the narrative of the West, manifesting in both the tangible paths they created and the profound significance of their pioneering spirit.
1. Impact on the Movement Westward

The mountain men, among the earliest non-Indigenous settlers of the Rocky Mountains and adjacent regions, played a crucial role in facilitating the movement toward westward expansion. In the early 19th century, a significant portion of the western United States was still shrouded in mystery for Euro-American settlers. The thorough explorations undertaken by the mountain men significantly enhanced cartographic accuracy, imparting essential geographical insights that would subsequently facilitate migration, settlement, and military endeavors.

The mountain men ventured into the newly acquired lands of the Louisiana Purchase, traversed the Rocky Mountain passes, and mapped rivers that would eventually become vital arteries for transportation. Their profound knowledge of the terrain—the mountains, lowlands, rivers, and natural obstacles—allowed them to serve as navigators for the settlers and explorers who came after them. Individuals such as Jedidiah Smith and Jim Bridger provided intricate insights into the western terrain, which were essential in the expansion of the United States' frontiers.

Beyond their adventurous pursuits, the mountain men significantly contributed to the economic growth of the West. Their main emphasis lay in the fur trade, yet their engagements with Indigenous communities, trading posts, and settlers played a crucial role in the formation of early trade networks. The networks in question enabled the transfer of goods, ideas, and cultural practices, thereby establishing a crucial groundwork for economic endeavors within the region.

2. The Evolution of Pathways and Routes of Exploration

The tangible heritage of the mountain men is most clearly illustrated in the trails and routes they forged, numerous of which evolved into essential corridors for westward expansion. In the course of their fur-trapping expeditions, mountain men explored uncharted territories, uncovering passes that would eventually be integrated into the Oregon, California, and Santa Fe Trails. These pathways would lead countless pioneers journeying westward in pursuit of fresh prospects during the mid-19th century.

Jim Bridger, celebrated as a prominent figure of the American frontier, played a pivotal role in discovering and mapping numerous essential routes, notably Bridger's Pass, which subsequently became integral to the Overland Trail. His role as a guide for emigrant wagon trains and military expeditions played a crucial part in forging essential links between the eastern United States and the frontier.

Jedidiah Smith's exploration of the Sierra Nevada and the Great Basin played a significant role in shaping the development of western routes. His extensive journeys offered revelations about regions that had remained obscure to settlers, facilitating the transportation of individuals and commodities through challenging landscapes. Smith's endeavors to map the West Coast and link the Rocky Mountains with California serve as a testament to the lasting impact of the mountain men on the region's geography and infrastructure.

The trails forged by the mountain men transcended mere pathways in the wild; they served as vital connections between diverse areas, facilitating the movement of settlers, the exchange of goods, and ultimately the founding of towns and cities. The westward expansion of the United States would have faced significantly greater challenges and delays without the groundbreaking contributions of these resilient individuals.

B. Cultural Archetype

In addition to their concrete impacts on the evolution of the American West, the mountain men bequeathed a significant cultural heritage. Their experiences and narratives contributed significantly to the formation of the archetype of the rugged individualist, a character that has become profoundly ingrained in American folklore, literature, and media.

1. The Significance of Freedom, Resilience, and the Wild Essence

The mountain men are frequently depicted as prime examples of autonomy, self-sufficiency, and fortitude. They existed in a manner characterized by isolation, exploration, and the resilience to endure the most unforgiving environments. The portrayal of the mountain man as a robust and wild character struck a profound chord within the American psyche, especially in an era when the frontier was perceived as a realm of possibility and liberation.

Their bond with the wilderness, characterized by both skill and reverence, represented a distinctive link to the natural realm. The mountain men flourished in a landscape that was as stunning as it was harsh, confronting its trials with bravery and resourcefulness. This embodiment of the mountain man transformed into a symbol of the American spirit of individualism and the quest for freedom, motivating countless generations to adopt the essence of exploration and self-reliance.

This image, frequently idealized, embodies a lasting intrigue with the concept of the wild West as a realm where human capability and determination could be challenged and realized. The legacy of the mountain men endures, evoking not just admiration but also a deep nostalgia for an era when the frontier symbolized boundless opportunity and significant trials.

## 2. Impact on American Folklore, Literature, and Media

The narratives of the mountain men have been enshrined in the annals of American folklore, profoundly influencing the cultural identity of the nation and its perception of the frontier. Stories of iconic individuals such as Jim Bridger, Jedidiah Smith, and Hugh Glass have been transmitted across generations, frequently enhanced to highlight their bravery, resilience, and extraordinary adventures.

These narratives have woven themselves into the fabric of literature, where they have been both honored and scrutinized as integral components of the expansive tale of the American West. Authors like James Fenimore Cooper and Mark Twain found their muse in the frontier experience, intricately crafting narratives that explore themes of wilderness, exploration, and survival. The mountain men, characterized by their audacious exploits and unyielding resolve, emerged as pivotal characters within this literary heritage.

Throughout the 20th and 21st centuries, the enduring legacy of the mountain men has found a vibrant expression in popular media. Motion pictures like The Revenant, which draws inspiration from the tale of Hugh Glass, alongside literary works such as Vardis Fisher's Mountain Men, have introduced these narratives to fresh audiences, igniting the curiosity of both viewers and readers alike. These representations frequently intertwine factual history with imaginative components, enhancing the legendary reputation of the mountain men while simultaneously delving into the intricacies of their existence.

The mountain men have emerged as lasting icons within the realms of art, music, and visual culture. Artworks by figures like Alfred Jacob Miller and Charles Marion Russell illustrate their experiences and surroundings, honoring the allure and peril of the frontier. Melodies and lyrical narratives have been crafted to honor their exploits, thereby solidifying their significance within the collective memory of society.

# THE END

www.ingramcontent.com/pod-product-compliance
Lightning Source LLC
Chambersburg PA
CBHW071203140325
23506CB00013B/1039

* 9 7 8 1 3 0 0 5 2 4 1 2 0 *